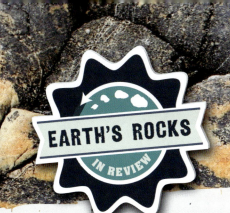

METAMORPHIC ROCKS

By Anna McDougal

Enslow PUBLISHING

Please visit our website, www.enslow.com. For a free color catalog of all our high-quality books, call toll free 1-800-398-2504 or fax 1-877-980-4454.

Cataloging-in-Publication Data
Names: McDougal, Anna.
Title: Metamorphic rocks / Anna McDougal.
Description: New York : Enslow Publishing, 2024. | Series: Earth's rocks in review | Includes glossary and index.
Identifiers: ISBN 9781978537910 (pbk.) | ISBN 9781978537927 (library bound) | ISBN 9781978537934 (ebook)
Subjects: LCSH: Metamorphic rocks–Juvenile literature.
Classification: LCC QE475.A2 M425 2024 | DDC 552'.4–dc23

Published in 2024 by
Enslow Publishing
2544 Clinton Street
Buffalo, NY 14224

Copyright © 2024 Enslow Publishing

Portions of this work were originally authored by Kristen Rajczak Nelson and published as *What Are Metamorphic Rocks?* All new material in this edition authored by Anna McDougal.

Designer: Claire Wrazin
Editor: Caitie McAneney

Photo credits: Cover, p. 1 Alex Comerford/Shutterstock.com; series art (title & heading background shape) cddesign.co/Shutterstock.com; series art (dark stone background) Somchai kong/Shutterstock.com; series art (white stone header background) Madredus/Shutterstock.com; series art (light stone background) hlinjue/Shutterstock.com; series art (learn more stone background) MaraZe/Shutterstock.com; p. 5 Guillermo Guerao Serra/Shutterstock.com; pp. 5, 15 arrows Elina Li/Shutterstock.com; p. 7 (top) Tyler Boyes/Shutterstock.com; pp. 7 (bottom), 9 (top), 25 (top & bottom), 29 (bottom) Yes058 Montree Nanta/Shutterstock.com; p. 9 (bottom) Aleksandr Pobedimskiy/Shutterstock.com; p. 11 meunierd/Shutterstock.com; p. 13 Nigel Jarvis/Shutterstock.com; p. 15 RealityImages/Shutterstock.com; p. 17 Ammit Jack/Shutterstock.com; p. 19 Amit kg/Shutterstock.com; p. 21 Islamic Footage/Shutterstock.com; p. 23 Pornpimon Ainkaew/Shutterstock.com; p. 27 JUN3/Shutterstock.com; p. 29 (top) LEONARDO VITI/Shutterstock.com.

All rights reserved. No part of this book may be reproduced in any form without permission in writing from the publisher, except by a reviewer.

Printed in the United States of America

Some of the images in this book illustrate individuals who are models. The depictions do not imply actual situations or events.

CPSIA compliance information: Batch #CWENS24: For further information, contact Enslow Publishing at 1-800-398-2504.

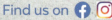

Changing Form 4
Metamorphic Makeup 6
Under Pressure 10
The Effects of Stress 12
Working Together 14
Finding Metamorphic Rocks 16
Regional Metamorphism 18
Contact Metamorphism 20
Identifying Rocks 22
Loads of Layers! 26
A Rock's Journey 28
Metamorphic Forces 30
Glossary 31
For More Information 32
Index 32

Words in the glossary appear in **bold** the first time they are used in the text.

CHANGING FORM

Earth is made up of rocks! One kind of rock is metamorphic rock. The word "metamorphic" comes from the Greek words *meta*, which means "change," and *morphe*, which means "form." These rocks start out as another kind of rock, then undergo a complete change.

hornfels (metamorphic rock)

LEARN MORE

The three kinds of rock on Earth are **igneous**, **sedimentary**, and metamorphic. Like sedimentary rocks, metamorphic rocks can be formed from existing rock.

METAMORPHIC MAKEUP

New conditions around a rock, such as heat or pressure, may cause metamorphism. That is the changing of a rock's **chemical** or **physical** composition, or makeup. The rock's composition changes in order to remain **stable** in the new conditions.

LEARN MORE

Pressure is a force that pushes on something else. Heat and pressure make the sedimentary rock shale into the metamorphic rock slate.

shale

slate

Rocks are made up of minerals. A mineral is matter that has a certain chemical makeup and an orderly structure. Minerals form rocks. When conditions change, the minerals in the rock **react** in different ways and take on different forms.

LEARN MORE

Marble is a metamorphic rock formed when limestone changes due to heat and pressure.

marble

limestone

UNDER PRESSURE

The deeper into Earth a rock is pushed, the more heat and pressure act on it. These two conditions often work together to form metamorphic rock. Pressure is caused by the weight of the rocks above and around the rock undergoing a change.

rock curved by pressure

LEARN MORE

Fluids, or liquids in a rock, can play a part in forming metamorphic rock. Matter carried by a fluid can flow into rock causing chemical changes to happen.

THE EFFECTS OF STRESS

Stress is a condition that causes a rock to change. When force is applied to one or two sides of a rock, it's called stress. If the stress makes the rock change shape, it's called strain.

LEARN MORE

Folds in rock are one example of how stress can affect rock.

WORKING TOGETHER

The forces that can cause metamorphism often work together to form metamorphic rock. For example, heat, pressure, and fluids can team up to change a rock. However, they can also cause metamorphism on their own.

gneiss

LEARN MORE

Gneiss is one kind of metamorphic rock that can form when the igneous rock granite comes under great heat and pressure.

15

FINDING METAMORPHIC ROCKS

Have you ever felt the earth shake? Earthquakes happen in places where tectonic plates meet. These large, rocky pieces of Earth's **crust** crash together, slide past one another, or move over or under each other. That's where you'll find metamorphic rock!

volcano

LEARN MORE

Volcanoes also occur where tectonic plates meet. Coming into contact with hot, liquid rock within volcanoes can cause rocks to change into metamorphic rocks.

17

REGIONAL METAMORPHISM

Many mountain ranges are made up of rocks formed by regional metamorphism. This happens over a large area when tectonic plates or other bodies of rock move. They lift rock up or push it deeper into Earth as they crash into one another or slide past each other.

LEARN MORE

When tectonic plates slide or crash, temperature and pressure commonly increase, and stress and strain may occur.

CONTACT METAMORPHISM

While regional metamorphism happens over a large area, contact metamorphism happens in a small area. It can occur when hot, liquid rock from within Earth flows into or near solid rock. It raises the temperature of the solid rock enough to cause changes.

LEARN MORE

Hot liquid rock is called lava when it's on Earth's surface or magma when it's underneath the surface.

IDENTIFYING ROCKS

People identify rocks by their **features**. One important feature is texture. When talking about rocks, "texture" means the size of a mineral's **crystals** or pieces called grains in a rock. Some rocks have fine, or small, grains. Others have coarse, or large, grains.

LEARN MORE

Scientists identify minerals in rocks by their hardness, color, and **cleavage**.

23

Metamorphic rock can be found with two different textures, foliated and nonfoliated. Foliated rocks have many minerals you can see in layers or bands. The minerals in nonfoliated rocks do not look aligned, or like they line up.

LEARN MORE

Nonfoliated rocks are formed under more uniform, or even, pressure than foliated rocks.

gneiss (foliated)

quartzite (nonfoliated)

LOADS OF LAYERS!

One kind of foliated rock with many layers is schist. Schist includes any foliated rock with large, flat minerals in thin layers. These rocks often have the word "schist" in their name, such as mica schist or garnet schist.

mica schist

LEARN MORE
Schist can form from any kind of rock and more than one kind of metamorphism.

A ROCK'S JOURNEY

Metamorphic rocks are often hard to identify. That's because two metamorphic rocks may look similar, but have little else in common. They may have had completely different parent rocks or reactions to new conditions. Every metamorphic rock has its own story!

LEARN MORE

Quartzite and marble often look alike. However, they are different metamorphic rocks made of different minerals.

PRESSURE

HEAT

METAMORPHIC FORCES

FLUIDS

STRESS

STRAIN

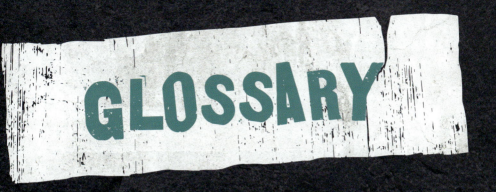

GLOSSARY

chemical: Matter that can be mixed with other matter to cause changes.

cleavage: How often a mineral breaks along smooth planes.

crust: The outer shell of a planet.

crystal: A hard piece of a substance formed when the substance turns into a solid. It often has many sides.

feature: An interesting or important part, look, or way of being.

igneous: Having to do with the rock that forms when hot, liquid rock from within Earth rises and cools.

physical: Having to do with the form of something.

react: To respond in a certain way.

sedimentary: Having to do with the rock that forms when sand, stones, and other matter are pressed together over a long time.

stable: Not likely to change suddenly or greatly.

volcano: An opening in Earth's surface through which hot, liquid rock sometimes flows.

BOOKS

Owen, Ruth. *Metamorphic Rocks.* Minneapolis, MN: Bearport Publishing, 2022.

Rogers, Marie. *Exploring the Rock Cycle.* New York, NY: PowerKids Press, 2022.

WEBSITE

How Are Rocks Made?
wonderopolis.org/wonder/how-are-rocks-made
Discover how rocks are made through a simple lesson and activities!

Publisher's note to educators and parents: Our editors have carefully reviewed this website to ensure it is suitable for students. Many websites change frequently, however, and we cannot guarantee that a site's future contents will continue to meet our high standards of quality and educational value. Be advised that students should be closely supervised whenever they access the internet.

chemical change, 11

crystal, 22

fluids, 11, 14, 30

heat, 6, 8, 10, 14, 15, 19, 20, 30

igneous rock, 5, 15

pressure, 6, 8, 10, 15, 19, 24, 30

sedimentary rock, 5, 6

strain, 12, 13, 19

stress, 12, 13, 19

tectonic plates, 16, 17, 18, 19

texture, 22, 23

volcanoes, 17